ABOUT ME	3
Formula, Which I pull out from Fermat Last Theorem	9
Multiplication is addition	13
Ancient Egyptian Multiplication and Fermat Last Theorem	17
Collatz conjecture attempt	22
Additive 50 - 51 Rupee puzzle (Solution theory)	35

April 2020

About Me

My name is Raheel H. Bhatti. I completed my Master Degree in Computer Science in Georgian-Tbilisi (Ex-USSR State) in the period from 1983-1988. I also studied Economics and Statistics.

During my study period I took interest in "**Fermat Last Theorem**".

$x^n + y^n = z^n$

Where n > 2 and x, y and z are Integers.

I informed to my Lecturer Mr. Ivan I. Chakhaidze. He looked at me a few seconds and then questioned "Are you sure?". I was speechless and answered, "With all due respect, I will try." He smiled and said. "Raheel, I know many mathematicians who lost their whole life in solving this formula. Try it but don't drive crazy. Give more time to your thesis". His statement forced me to think twice on this matter and I decided to spend one to two hours daily on **Fermat Last Theorem**.

I am not a Mathematician but I have a great interest in Number Theory. My thesis was "Recursive Functions" and wrote my thesis on a virtual machine "Machine with Unlimited Registers" (MUR). This machine is not exact but like "Turing- Machine". In MUR I learned Addition, Subtraction, and

Division but I realized that there is no Program for multiplication. I decided to write a program on multiplication. After struggling I wrote a multiplication program and also a program to convert decimal to binary and vice versa in MUR and then inform to my lecturer. He examined the Programs and suggested me to participate in the "State Georgian Student Science Conference for Programming 1987-1988".

After few months held the Conference and I presented my multiplication program. (The root of this program is to save the time). I was awarded with a "Winner-Diploma" of this Conference.

In the meantime, I also worked on **Fermat Last Theorem**. To understand deeply I started to strip the powers. I wrote down $1^3, 2^3, 3^3, 4^3, 5^3...$ and subtracted the first value from the second value and the second value from the third value and so on. I did it three times and get a constant number "6". 6 is the factorial of 3. For me it was amazing. I did it for the fourth power and get "24" which is the factorial of 4. I thought I have something in my hand. This gave me more motivation and I pulled out another Formula.

I informed again to my Lecturer. He said to me that "I am not mathematician please go to the Academy of Mathematics" in Tbilisi.

I wrote a letter and got an appointment to present my new Formulas. I went there and under three observers I presented my formulas. First, I detailed them about the subtractions of values and to get a constant factorial number of the power. **They smiled**! They appreciated my struggle and told me that Isaac Newton already found. (As I mentioned before I am not a mathematician). I was unknowledge on Isaac Newton's work.

I wrote down my second formula on blackboard. The looked-for few minutes on the blackboard and told me "This is something new". One of them said that it also satisfied in negative numbers. After half an hour of discussion on this Formula they gifted me some books on "Number Theory" and "**Format Last Theorem**". (See this Formula in the section "Formula, which I pull out from Fermat Last Theorem".).

I didn't Published my Formula as I wanted to gain more from this Formula.

Since 1988 I am living in Germany. I got a job as a programmer and took no interest in the Formula but little. Now I realize that if I work further on it may take more time. I decide to publish a Book.

Formula, Which I pull out from Fermat Last Theorem

$a^3 + b^3 \neq c^3$

First, we take the power of three. As we know:

$9^3 + 10^3 \neq 12^3$

$$729 + 1000 = 1728 + \boxed{1}$$

$1 = 10 - 9$

$1 = 1^3$

Now we have the equation:

$9^3 + 10^3 = 12^3 + (10 - 9)^3$

729 + 1000 = 1728 + 1

Here I wanted to get a sequence that can satisfy the equation. I wrote down from 1^3 to 100^3 in a row to find the next numbers.

Result:

$18^3 + 20^3 = 24^3 + (20 - 18)^3$

$27^3 + 30^3 = 36^3 + (30 - 27)^3$

$36^3 + 40^3 = 48^3 + (40 - 36)^3$

Now we have sequence and Formula.

$$(ak)^3 + (bk)^3 = (ck)^3 + (bk - ak)^3$$

Where,

a = 9

b = 10

c = 12

k = 1, 2, 3, 4, 5, 6, 7, 8, ….

In simple words we write the following sequence:

$$9^3 + 10^3 = 12^3 + 1^3$$

$$18^3 + 20^3 = 24^3 + 2^3$$

$$27^3 + 30^3 = 36^3 + 3^3$$

$$36^3 + 40^3 = 48^3 + 4^3$$

$$45^3 + 50^3 = 60^3 + 5^3$$

And so on.

Multiplication is addtion

Multiplication is an addition. If we multiply 2 factors, it means that we add the first factor, second factor times.

For example:

5.4 = 20 or

5+5+5+5 = 20

We add 4 times the 5.

a^n multiplying b^n is alwyas equal to c^n

Where c = a x b

Formula:

$a^n . b^n = c^n$

where c = a x b

Summing the constant:

a^n is a constant. Add a^n b^n times is equal to the c^n where c = a.b

$$\sum_{k=a^n}^{d} s = kd$$

d = $(b)^n - 1$

we will add k (a^n) d times.

Examples:

Let n = 3

a = 3 and b = 7 then c = 21

$3^3 . 7^3 = 21^3$

Where 21 = 3.7

$\underline{27.343 = 9261}$

Or add 27 times 343.

$6^4 \cdot 4^4 = 24^4$

Where 24 = 4.6

$6^4 = 1296$

$4^4 = 256$ and

$24^4 = 331776$

$\underline{1296.256 = 331776}$

$2^6 \cdot 4^6 = 8^6$

Where 8 = 2.4

$2^6 = 64$

$4^6 =$ 4096

$8^6 =$ 262144

64.4096 = 262144

$3^7 . 2^7 = 6^7$

Where 6 = 3.2

$3^7 =$ 2187

$2^7 =$ 128

$6^7 =$ 279936

2187.128= 279936

Ancient Egyptian Multiplication and Fermat Last Theorem

Ancient Egyptian Multiplication:

Just take any two integers and put them into two columns. Better to put the small integer in the first column and bigger in the second. Many of you know the ancient Egyptian multiplication. Increase the value in first from 1 and double it till it does not accede the integer in the first column. Second column integer also double till it reaches to the last row of the first column. (See table)

17	27
①	㉗
2	54
4	108
8	216
⑯	㊵㉜

We see that the sum of the 16 and 1 give us the integer not acceding to the first column value.

17 x 27 = 459

27 + 432 = 459.

As we now from previous chapter that $a^n.b^n = c^n \text{ where } c = a.b$

Now is interesting.

$27^n + 432^n / 1^n + 16^n = 27^n$

Let we take another example:

1st Column integer = 25

2nd Column integer = 72

25	72
①	⑦②
2	144
4	288
⑧	⑤⑦⑥
⑯	⑪⑮②

16+8+1 = 25

72+576+1152 = 25x72= 1800

$72^n + 576^n + 1152^n / 1^n + 8^n + 16^n = 72^n$

$72^3 + 576^3 + 1152^3 / 1^3 + 8^3 + 16^3 = 72^3$

Now we take an example of two little big integers as 235 x 454.

235 in the first column and 454 in the second column.

235	454
①	⟨454⟩
②	⟨908⟩
4	1816
⑧	⟨3632⟩
16	7264
㉜	⟨14528⟩
㊿... ⟨64⟩	⟨29056⟩
⟨128⟩	⟨58112⟩

235 x 454 = 106690 ◄

Sum of the selected vales from column two:

454+908+3632+14528+29056+58112=⟨106690⟩

The sum of the individual selected $(Value)^n$ from column two divided by the sum of the individual selected $(Value)^n$ from column one is equal to the power of n of the multiplicand i.e. 454.

As shown below:

$$\frac{454^n+908^n+3632^n+14528^n+29056^n+58112^n}{1^n+2^n+8^n+32^n+64^n+128^n}$$

$= 454^n$

Let n = 3

$$\frac{454^3+908^3+3632^3+14528^3+29056^3+58112^3}{1^3+2^3+8^3+32^3+64^3+128^3}$$

$= 454^3$

Collatz Conjecture

Ursprung und Geschichte [Bearbeiten | Quelltext bearbeiten]

Der Ursprung der Collatz-Vermutung liegt insofern etwas im Nebel, als aus der mutmaßlichen Entstehungszeit bisher keine schriftlichen Dokumente mit einer Beschreibung des Problems öffentlich zugänglich sind. Es wird berichtet, dass Collatz das Problem beim Internationalen Mathematikerkongress 1950 in Cambridge (Massachusetts) mündlich verbreitete.[10] Stanisław

Source:

https://de.wikipedia.org/wiki/collatz-problem

Translation in English:

Origin and history

The origin of the Collatz conjecture lies somewhat in the fog, since so far no written documents describing the problem are publicly available from the supposed time of origin. Collatz is reported to have spoken orally at the 1950 International Congress of Mathematicians in Cambridge, Massachusetts.

According to the source from Wikipedia (as above) "***The origin of the Collatz conjecture lies somewhat in the fog***" & "***Collatz is reported to have spoken orally***". Now we take the Collatz conjecture theory under the loop.

As we know from different sources from the internet and YouTube clips on Collatz conjecture, if n is even, then n divided by 2 (n/2) and if n reach to an odd number then multiply n with three and add 1 to the result (3(n) + 1).

Now let us try to solve Collatz-conjecture with another method.

Collatz Conjecture

Let we take n = 5

5 is an odd number so according to rules 5 will multiply with 3 and add 1 to the result.

n = (3(5) + 1)

n = 16

Now we divide the 16 by 2 as 16 is an even number. After division we get the result 8. We have also rest which is 8 as 8+8=16. We will not dispose of the rest as which is half of the 16. We use the 8 for our further calculation, we will also calculate the other half 8. (8/2) will give us 4. 8 have two parts 4 and 4 and they make 8. The other 8 will also be divided by 2 and give us two times 4. Sum of the 4 times 4 is 16, our origin even number (16).

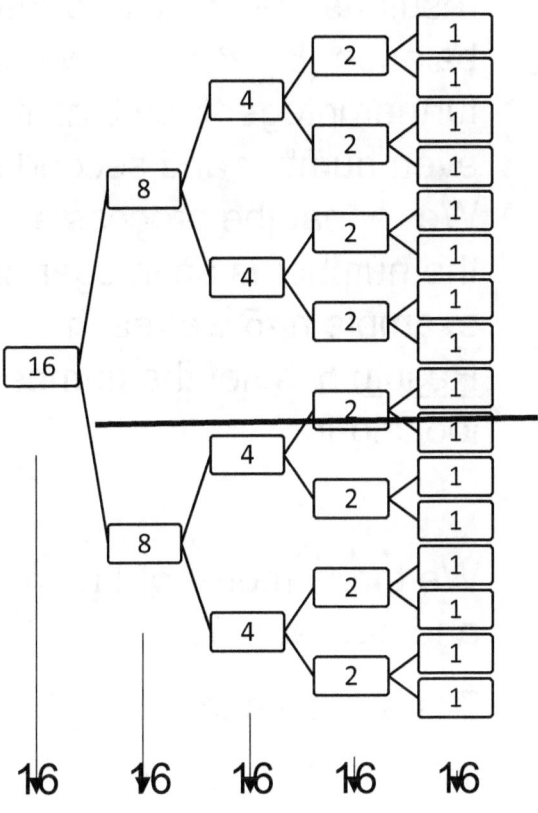

We divide the integer 16 (from integer 5, because integer 5 is an odd number, we multiply with 3 add

1 to the result) and get 8. Here we are not disposing the rest 8 because this will give us the same result as the first 8 and maintain the balance (8 + 8 = 16). we divide further integer 8 by 2 as it is an even number and second 8 also. We repeat the process as long as the number is an integer. In this example n=5 we reach on 1.

<u>Result</u>: 5 is not the number we are looking for.

We throw more light to the theory and say let n=7.

7 is an odd number: 3(7) +1=22

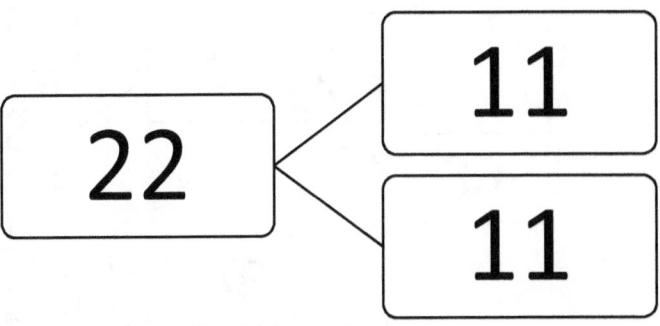

In our last example we get an even number 8 by dividing 16 by 2. We also divide the both 8s and four times 4. But here we have a deal now with an odd number. According to the rule odd number will multiply with 3 and add 1 to the result. We will do it with both 11s. it means 2(11(3) +1) n= 68.

68 is greater than 22 (68 > 22 or 68 ≠ 22). We have now a new integer, so we start from 68.

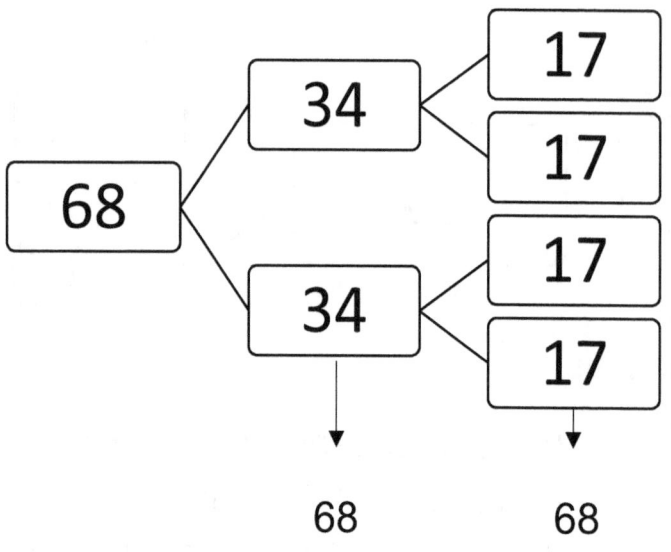

17 is an odd number. According to rule, odd numbers will multiply with 3 and we add 1 to the result. We are not disposing the rest integer as they are the part of the master integer and the same process is running for the rest integers also as for the first. It means…

4(3(17) + 1) = 208 is new integer and not equal to master. 208 > 68 or 208 ≠ 68.

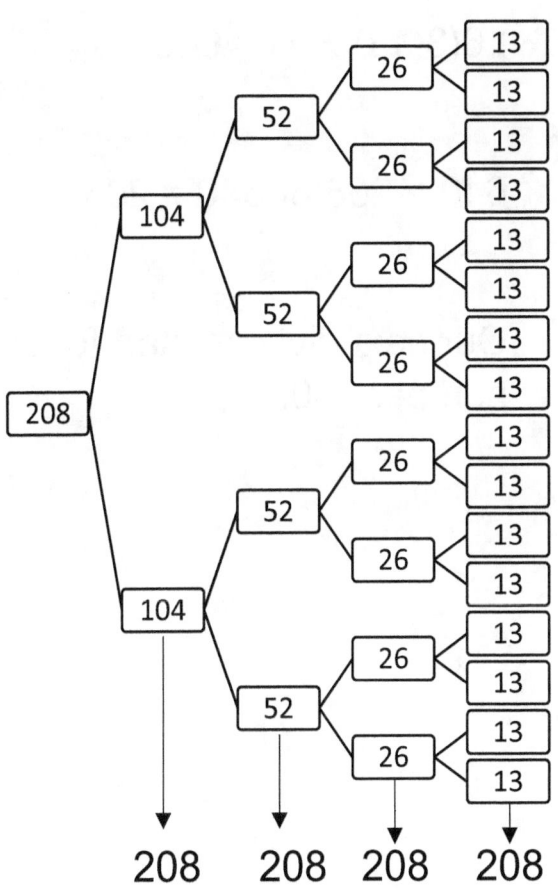

208 division reach on 13. 13 is now odd number and 16 times makes our master integer 208.

$$16(3(13) + 1) = 640$$

640 > 208 or 640 ≠ 208

Our next division start for even number 640.

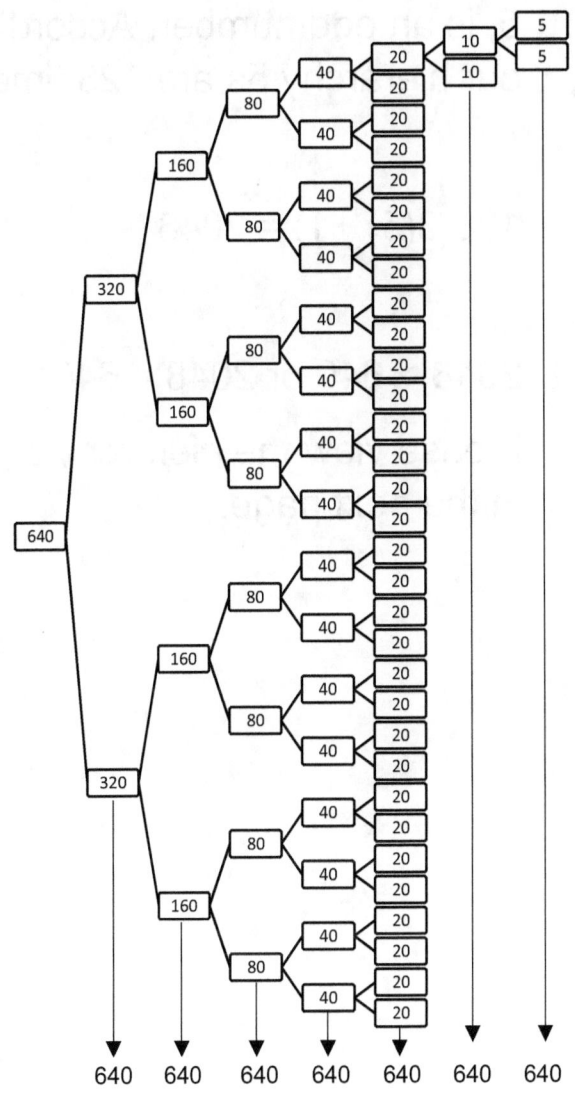

5, is an odd number. According to our hierarchy 5s are 128 times.

$128(3(5) +1) = 2048$

$2048 > 640$ or $2048 \neq 640$

Please view the hierarchy of 2048 in the next page.

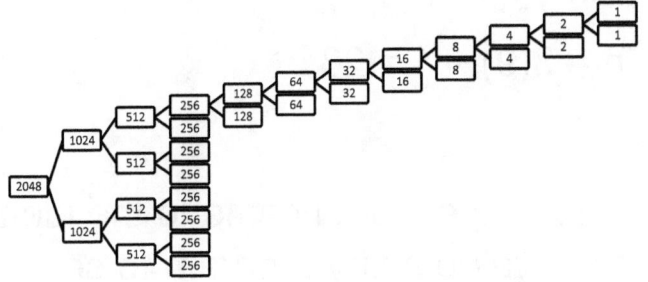

7 reach to 1. 7 is not our integer. (for saving and clear view purpose I short the hierarchy)

Next integer 9

n = 9 an odd number. According to rule:

n = 3(9) + 1 = 28

After paperwork I came to the idea to develop a tiny program to see the results. Every given integer ended on 4-2-1.

Is Collatz conjecture UNCRACKABLE?

Additive
50 Rupee puzzle, 1 Rupee balance error solution.

1 Rupee balance problem is designated as toughest Riddle – where did the Rupee 1 come from?

I will try my best to explain the hidden philosophy of the 1 Rupee. Here is the philosophy of seen and unseen. Seen and unseen are not the same.

Here 50 Rupee in hand is seen. According to the Question a person spends 20 Rupee and left 30. After a while he/she spends 15 Rupee and in balance is now 15. He/she spends further Rupees 9 and now in the balance sheet is 6 Rupees. After spending the rest 6 Rupees now balance is 0. If we calculate the Spend column, it is 50 but the sum of the Balance column is 51.

Rs. 50/-

Spend	Balance
20	30
15	15
9	6
6	0
50	51

Why it is so?

The real money, what a person has in hand is counted from 1 to 50.

Balance is counted from 0 to 50. You can also say real numbers are counted from 1 to x and balance numbers are from 0 to x.

Prove:

Balance = 51

Spend = 50

x = Balance − Spend

x = 51 − 50 = 1

Now we make the 50 parts of the 1.

y = 1 / 50 = 0.02

Every time when we spend, we calculate the 50th part of the spend amount. In our example we spend the first 20 Rupees. It means:

z_1 = 20 x 0.02 = 0.4

In reality we have 30 Rupee in hand but in the Real Balance Column we have 29.60 Rupees.

Balanced 30 Rupees – z_1 = 29.40

We spend further 15 Rupees. Real money in our hand remains 15 but in Balance column we will subtract 0.3 (z) it is part of the money we spend.

$z_2 = 15 \times 0,02$

15 – z_2 = 14.70

We do the same for every row.

z_3 = 9 x 0.02 = 0.18 (9 – z_3 = 5.82)

z_4 = 6 x 0.02 = 0.12 (0 - z_4 = 0.12)

Please see the table.

Rupee 50.00

Spend	Balance	Real Balance
20	30	29.60
15	15	14.70
9	6	5.82
6	0	-0.12
50	51	50

Example 1:

x = Sum of balance − sum of spend

$x = 10 - 50 = -40$

$y = -40/50 = -0.80$

Spend 40:

$40 \times -0.80 = -32$

Spend 10:

$10 \times -0.80 = -8$

Rupee 50.

Spend	Balance	Real Balance
40	10	10 – (-32) =42
10	0	0 – (-8) = 8
50	10	50

Example 2:

Rupee 50.00

Spend	Balance	Real Balance
10	40	30
10	30	20
10	20	10
10	10	0
10	0	-10
50	100	50

$100 - 50 = 50$

$50/50 = 1$

Conclusion:

After spending whole money.

x = sum of Balance – sum of Spend

$y = x/$sum of spend column

$z1$ = Spend1(y)

$z2$ = Spend2(y)

….

zn = Spendn(y)

rb1 = Real Balance of row 1

rb1 = Balance1 – z …

$$\sum_{z=i}^{n} zi = \sum_{rb=i}^{n} rbi$$

I am sorry if I tipped wrong or is explained in not a proper way. I tried my best to explain the modules in an easy way.

For any detail or questions please contact me on the following e-mail.

bhattiraheel@t-online.de

© Raheel Hasan Bhatti

www.ingramcontent.com/pod-product-compliance
Lightning Source LLC
Chambersburg PA
CBHW050305220526
45465CB00002B/836